'The vessel
drove before
her bows two
billows of liquid
phosphorus . . .'

CHARLES DARWIN
Born 1809, Shrewsbury, Shropshire
Died 1882, Downe, Kent

Extracts taken from *The Voyage of the Beagle*,
first published in 1839.

CHARLES DARWIN IN PENGUIN CLASSICS
Autobiographies
On the Origin of Species
The Descent of Man
The Expression of the Emotions in Man and Animals
The Voyage of the Beagle

CHARLES DARWIN

It was snowing butterflies

PENGUIN BOOKS

PENGUIN CLASSICS

Published by the Penguin Group
Penguin Books Ltd, 80 Strand, London WC2R ORL, England
Penguin Group (USA) Inc., 375 Hudson Street, New York, New York 10014, USA
Penguin Group (Canada), 90 Eglinton Avenue East, Suite 700, Toronto, Ontario,
Canada M4P 2Y3 (a division of Pearson Penguin Canada Inc.)
Penguin Ireland, 25 St Stephen's Green, Dublin 2, Ireland
(a division of Penguin Books Ltd)
Penguin Group (Australia), 707 Collins Street, Melbourne, Victoria 3008, Australia
(a division of Pearson Australia Group Pty Ltd)
Penguin Books India Pvt Ltd, 11 Community Centre, Panchsheel Park,
New Delhi – 110 017, India
Penguin Group (NZ), 67 Apollo Drive, Rosedale, Auckland 0632, New Zealand
(a division of Pearson New Zealand Ltd)
Penguin Books (South Africa) (Pty) Ltd, Block D, Rosebank Office Park,
181 Jan Smuts Avenue, Parktown North, Gauteng 2193, South Africa

Penguin Books Ltd, Registered Offices: 80 Strand, London WC2R ORL, England

www.penguin.com

This selection published in Penguin Classics 2015
001

Set in 9.5/13 pt Baskerville 10 Pro
Typeset by Jouve (UK), Milton Keynes
Printed in Great Britain by Clays Ltd, St Ives plc

A CIP catalogue record for this book is available from the British Library

ISBN: 978-0-141-39855-6

www.greenpenguin.co.uk

FSC
www.fsc.org

MIX
Paper from
responsible sources
FSC™ C018179

Penguin Books is committed to a sustainable
future for our business, our readers and our planet.
This book is made from Forest Stewardship
Council™ certified paper.

Contents

Patagonia

DECEMBER 6TH, 1833 – The *Beagle* sailed from the Rio Plata, never again to enter its muddy stream. Our course was directed to Port Desire, on the coast of Patagonia. Before proceeding any further, I will here put together a few observations made at sea.

Several times when the ship has been some miles off the mouth of the Plata, and at other times when off the shores of Northern Patagonia, we have been surrounded by insects. One evening, when we were about ten miles from the Bay of San Blas, vast numbers of butterflies, in bands or flocks of countless myriads, extended as far as the eye could range. Even by the aid of a glass it was not possible to see a space free from butterflies. The seamen cried out 'it was snowing butterflies', and such in fact was the appearance. More species than one were present, but the main part belonged to a kind very similar to, but not identical with, the common English *Colias edusa*. Some moths and hymenoptera accompanied the butterflies; and a fine Calosoma flew on board. Other instances are known of this beetle having been caught far out at sea; and this

1

is the more remarkable, as the greater number of the Carabidæ seldom or never take wing. The day had been fine and calm, and the one previous to it equally so, with light and variable airs. Hence we cannot suppose that the insects were blown off the land, but we must conclude that they voluntarily took flight. The great bands of the Colias seem at first to afford an instance like those on record of the migrations of *Vanessa cardui*; but the presence of other insects makes the case distinct, and not so easily intelligible. Before sunset, a strong breeze sprung up from the north, and this must have been the cause of tens of thousands of the butterflies and other insects having perished.

On another occasion, when 17 miles off Cape Corrientes, I had a net overboard to catch pelagic animals. Upon drawing it up, to my surprise I found a considerable number of beetles in it, and although in the open sea, they did not appear much injured by the salt water. I lost some of the specimens, but those which I preserved, belonged to the genera, colymbetes, hydroporus, hydrobius (two species), notaphus, cynucus, adimonia, and scarabæus. At first, I thought that these insects had been blown from the shore; but upon reflecting that out of the eight species, four were aquatic, and two others partly so in their habits, it appeared to me most probable that they were floated into the sea, by a small stream which drains a lake near Cape Corrientes. On any supposition, it is an interesting circumstance to find insects, quite alive, swimming in the open ocean, 17 miles from the nearest point of land. There are several accounts of insects having been blown off the Patagonian shore. Captain Cook observed it, as did more lately Captain King in the *Adventure*. The cause probably is due to the want of shelter, both of trees and hills, so that an insect on the wing with an off-shore breeze, would be very apt to be blown out to sea. The most remarkable instance I ever knew of an insect being caught far from the land, was that of a large grasshopper (*Acrydium*), which flew on board, when the *Beagle* was to windward of the Cape de Verd Islands, and when the nearest point of land, not directly opposed to the trade-wind, was Cape Blanco on the coast of Africa, 370 miles distant.

On several occasions, when the vessel has been within the mouth of the Plata, the rigging has been coated with the web of the Gossamer Spider. One day (November 1st, 1832) I paid particular attention to the phenomenon. The weather had been fine and clear, and in the morning the air was full of patches of the flocculent web, as on an autumnal day in England. The ship was sixty miles distant from the land, in the direction of a steady though light breeze. Vast numbers of a small spider, about one-tenth of an inch in length, and of a dusky red colour were attached to the webs. There must have been, I should suppose, some thousands on the ship. The little spider when first coming in contact with the rigging, was always seated on a single thread, and not on the flocculent mass.

This latter seems merely to be produced by the entanglement of the single threads. The spiders were all of one species, but of both sexes, together with young ones. These latter were distinguished by their smaller size, and more dusky colour. I will not give the description of this spider, but merely state that it does not appear to me to be included in any of Latreille's genera. The little aeronaut as soon as it arrived on board, was very active, running about; sometimes letting itself fall, and then reascending the same thread; sometimes employing itself in making a small and very irregular mesh in the corners between the ropes. It could run with facility on the surface of water. When disturbed it lifted up its front legs, in the attitude of attention. On its first arrival it appeared very thirsty, and with exserted maxillæ drank eagerly of the fluid; this same circumstance has been observed by Strack: may it not be in consequence of the little insect having passed through a dry and rarefied atmosphere? Its stock of web seemed inexhaustible. While watching some that were suspended by a single thread, I several times observed that the slightest breath of air bore them away out of sight, in a horizontal line. On another occasion (25th) under similar circumstances, I repeatedly observed the same kind of small spider, either when placed, or having crawled, on some little eminence, elevate its abdomen, send forth a thread, and then sail away in a lateral course, but with a rapidity which was quite unaccountable. I thought I could perceive that the spider before performing the above preparatory steps, connected its legs together with the most delicate threads, but I am not sure, whether this observation is correct.

One day, at St Fe, I had a better opportunity of observing some similar facts. A spider which was about three-tenths of an inch in length, and which in its general appearance resembled a Citigrade (therefore quite different from the gossamer), while standing on the summit of a post, darted forth four or five threads from its spinners. These glittering in the sunshine, might be compared to rays of light; they were not, however, straight, but in undulations like a film of silk blown by the wind. They were more than a yard in length, and diverged in an ascending direction from the orifices. The spider then suddenly let go its hold, and was quickly borne out of sight. The day was hot and apparently quite calm; yet under such circumstances the atmosphere can never be so tranquil, as not to affect a vane so delicate as the thread of a spider's web. If during a warm day we look either at the shadow of any object cast on a bank, or over a level plain at a distant landmark, the effect of an ascending current of heated air will almost always be evident. And this probably would be sufficient to carry with it so light an object as the little spider on its thread. The circumstance of spiders of the same species but of different sexes and ages, being found on several occasions at the distance of many leagues from the land, attached in vast numbers to the lines, proves that they are the manufacturers of

5

the mesh, and that the habit of sailing through the air, is probably as characteristic of some tribe, as that of diving is of the Argyroneta. We may then reject Latreille's supposition, that the gossamer owes its origin to the webs of the young of several genera, as Epeira or Thomisa: although, as we have seen that the young of other spiders do possess the power of performing aerial voyages.

During our different passages south of the Plata, I often towed astern a net made of bunting, and thus caught many curious animals. The structure of the Beroe (a kind of jelly fish) is most extraordinary, with its rows of vibratory ciliæ, and complicated though irregular system of circulation. Of Crustacea, there were many strange and undescribed genera. One, which in some respects is allied to the Notopods (or those crabs which have their posterior legs placed almost on their backs, for the purpose of adhering to the under side of ledges), is very remarkable from the structure of its hind pair of legs. The penultimate joint, instead of being terminated by a simple claw, ends in three bristle-like appendages of dissimilar lengths, the longest equalling that of the entire leg. These claws are very thin, and are serrated with teeth of an excessive fineness, which are directed towards the base. The curved extremities are flattened, and on this part five most minute cups are placed, which seem to act in the same manner as the suckers on the arms of the cuttle-fish. As the animal lives in the open sea, and probably wants a place of rest, I suppose this beautiful structure is adapted to take hold of the globular bodies of the Medusæ, and other floating marine animals.

In deep water, far from the land, the number of living creatures is extremely small: south of the latitude 35°, I never succeeded in catching any thing besides some beroe, and a few species of minute crustacea belonging to the Entomostraca. In shoaler water, at the distance of a few miles from the coast, very many kinds of crustacea and some other animals were numerous, but only during the night. Between latitudes 56° and 57° south of Cape Horn the net was put astern several times; it never, however, brought up any thing besides a few of two extremely minute species of Entomostraca. Yet whales and seals, petrels and albatross, are exceedingly abundant throughout this part of the ocean. It has always been a source of mystery to me, on what the latter, which live far from the shore, can subsist. I presume the albatross, like the condor, is able to fast long; and that one good feast on the carcass of a putrid whale lasts for a long siege of hunger. It does not lessen the difficulty to say, they feed on fish; for on what can the fish feed? It often occurred to me, when observing how the waters of the central and intertropical parts of the Atlantic, swarmed with Pteropoda, Crustacea, and Radiata, and with their devourers the flying-fish, and again with *their* devourers the bonitos and albicores, that the lowest of these pelagic animals perhaps possess the power of decomposing carbonic acid gas, like the members of the vegetable kingdom.

While sailing in these latitudes on one very dark night, the sea presented a wonderful and most beautiful spectacle. There was a fresh breeze, and every part of the surface, which during the day is seen as foam, now glowed with a pale light. The vessel drove before her bows two billows of liquid phosphorus, and in her wake she was followed by a milky train. As far as the eye reached, the crest of every wave was bright, and the sky above the horizon, from the reflected glare of these livid flames, was not so utterly obscure, as over the rest of the heavens.

As we proceed further southward, the sea is seldom phosphorescent; and off Cape Horn, I do not recollect more than once having seen it so, and then it was far from being brilliant. This circumstance probably has a close connexion with the scarcity of organic beings in that part of the ocean. After the elaborate paper by Ehrenberg, on the phosphorescence of the sea, it is almost superfluous on my part to make any observations on the subject. I may however add, that the same torn and irregular particles of gelatinous matter, described by Ehrenberg, seem in the southern as well as in the northern hemisphere, to be the common cause of this phenomenon. The particles were so minute as easily to pass through fine gauze; yet many were distinctly visible by the naked eye. The water when placed in a tumbler and agitated gave out sparks, but a small portion in a watch-glass, scarcely ever was luminous. Ehrenberg states, that these particles all retain a certain degree of irritability. My observations, some of which were made directly after taking up the water, would give a different result. I may also mention, that having used the net during one night I allowed it to become partially dry, and having occasion twelve hours afterwards, to employ it again, I found the whole surface sparkled as brightly as when first taken out of the water. It does not appear probable in this case, that the particles could have remained so long alive. I remark also in my notes, that having kept a Medusa of the genus Dianæa, till it was dead, the water in which it was placed became luminous. When the waves scintillate with bright green sparks, I believe it is generally owing to minute crustacea. But there can be no doubt that very many other pelagic animals, when alive, are phosphorescent.

On two occasions I have observed the sea luminous at considerable depths beneath the surface. Near the mouth of the Plata some circular and oval patches, from 2 to 4 yards in diameter, and with defined outlines, shone with a steady, but pale light; while the surrounding water only gave out a few sparks. The appearance resembled the reflection of the moon, or some luminous body; for the edges were sinuous from the undulation of the surface. The ship, which drew thirteen feet water, passed over, without disturbing, these patches. Therefore we must suppose that some animals were congregated together at a greater depth than the bottom of the vessel.

Near Fernando Noronha the sea gave out light in flashes. The appearance was very similar to that which might be expected from a large fish moving rapidly through a luminous fluid. To this cause the sailors attributed it; at the time, however, I entertained some doubts, on account of the frequency and rapidity of the flashes. With respect to any general observations, I have already stated that the display is very much more common in warm than in cold countries. I have sometimes imagined that a disturbed electrical condition of the atmosphere was most favourable to its production. Certainly I think the sea is most luminous after a few days of more calm weather than ordinary, during which time it has swarmed with various animals. Observing that the water charged with gelatinous particles is in an impure state, and that the luminous appearance in all common cases is produced by the agitation of the fluid in contact with the atmosphere, I have always been inclined to consider that the phosphorescence was the result of the decomposition of the organic particles, by which process (one is tempted almost to call it a kind of respiration) the ocean becomes purified.

DECEMBER 23RD – We arrived at Port Desire, situated in lat. 47°, on the coast of Patagonia. The creek runs for about twenty miles inland, with an irregular width. The *Beagle* anchored a few miles within the entrance in front of the ruins of an old Spanish settlement.

The same evening I went on shore. The first landing in any new country is very interesting, and especially when, as in this case, the whole aspect bears the stamp of a marked and individual character. At the height of between 200 and 300 feet, above some masses of porphyry, a wide plain extends, which is truly characteristic of Patagonia. The surface is quite level, and is composed of well-rounded shingle mixed with a whitish earth. Here and there scattered tufts of brown wiry grass are supported, and still more rarely some low thorny bushes. The weather is dry and pleasant, for the fine blue sky is but seldom obscured. When standing in the middle of one of these desert plains, the view on one side is generally bounded by the escarpment of another plain, rather higher, but equally level and desolate; and on the other side it becomes indistinct from the trembling mirage which seems to rise from the heated surface.

The plains are traversed by many broad, flat-bottomed valleys, and in these the bushes grow rather more abundantly. The present drainage of the country is quite insufficient to excavate such large channels. In some of the valleys ancient stunted trees, growing in the very centre of the dry watercourse, seem as if placed to prove how long a time had elapsed, since any flood had passed that way. We have evidence, from shells lying on the surface, that the plains of gravel have been elevated within a recent epoch above the level of the sea; and we must look to that period for the excavation of the valleys by the slowly retiring waters. From the dryness of the climate, a man may walk for days together over these plains without finding a single drop of water. Even at the base

11

of the porphyry hills, there are only a few small wells containing but little water, and that rather saline and half putrid.

In such a country the fate of the Spanish settlement was soon decided; the dryness of the climate during the greater part of the year, and the occasional hostile attacks of the wandering Indians, compelled the colonists to desert their half-finished buildings. The style, however, in which they were commenced, showed the strong and liberal hand of Spain in the old time. The end of all the attempts to colonize this side of America south of 41°, have been miserable. At Port Famine, the name expresses the lingering and extreme sufferings of several hundred wretched people, of whom one alone survived to relate their misfortunes. At St Joseph's bay, on the coast of Patagonia, a small settlement was made; but during one Sunday the Indians made an attack and massacred the whole party, excepting two men, who were led captive many years among the wandering tribes. At the Rio Negro I conversed with one of these men, now in extreme old age.

The zoology of Patagonia is as limited as its Flora. On the arid plains a few black beetles (Heteromera) might be seen slowly crawling about, and occasionally a lizard darting from side to side. Of birds we have three carrion hawks, and in the valleys a few finches and insect feeders. The *Ibis malanops* (a species said to be found in central Africa) is not uncommon on the most desert parts. In the stomachs of these birds I found grasshoppers, cicadæ, small lizards, and even scorpions. At one time of the year they go in flocks, at another in pairs: their cry is very loud and singular, and resembles the neighing of the guanaco.

I will here give an account of this latter animal, which is very common, and is the characteristic quadruped of the plains of Patagonia. The Guanaco, which by some naturalists is considered as the same animal with the Llama, but in its wild state, is the South American representative of the camel of the East. In size it may be compared to an ass, mounted on taller legs, and with a very long neck. The guanaco abounds over the whole of the temperate parts of South America, from the wooded islands of Tierra del Fuego, through Patagonia, the hilly parts of La Plata, Chile, even to the Cordillera of Peru. Although preferring an elevated site, it yields in this respect to its near relative the Vicuna. On the plains of Southern Patagonia, we saw them in greater numbers than in any other part. Generally they go in small herds, from half a dozen to thirty together; but on the banks of the St Cruz we saw one herd which must have contained at least 500. On the northern shores of the Strait of Magellan they are also very numerous.

Generally the guanacoes are wild and extremely wary. Mr Stokes told me, that he one day saw through a glass a herd of these beasts, which evidently had been frightened, running away at full speed, although their distance was so great that they could not be distinguished by the naked eye. The sportsman frequently receives the first intimation of their presence, by hearing, from a long distance, the peculiar shrill neighing note of alarm. If he then looks attentively, he will perhaps see the herd standing in a line on the side of some distant hill. On approaching them, a few more squeals are given, and then off they set at an apparently slow, but really quick canter, along some narrow beaten track to a neighbouring hill. If, however, by chance he should abruptly meet a single animal, or several together, they will generally stand motionless, and intently gaze at him; then perhaps move on a few yards, turn round, and look again. What is the cause of this difference in their shyness? Do they mistake a man in the distance for their chief enemy the puma? Or does curiosity overcome their timidity? That they are curious is certain; for if a person lies on the ground, and plays strange antics, such as throwing up his feet in the air, they will almost always approach by degrees to reconnoitre him. It was an artifice that was repeatedly practised by our sportsmen with success, and it had moreover the advantage of allowing several shots to be fired, which were all taken as parts of the performance. On the mountains of Tierra del Fuego, and in other places, I have more than once seen a guanaco, on being approached, not only neigh and squeal, but prance and leap about in the most ridiculous manner, apparently in defiance as a challenge. These animals are very easily domesticated, and I have seen some thus kept near the houses, although at large on their native plains. They are in this state very bold, and readily attack a man, by striking him from behind with both knees. It is asserted, that the motive for these attacks is jealousy on account of their females. The wild guanacoes, however, have no idea of defence; even a single dog will secure one of these large animals, till the huntsman can come up. In many of their habits they are like sheep in a flock. Thus when they see men approaching in several directions on horseback, they soon became bewildered and know not which way to run. This greatly facilitates the Indian method of hunting, for they are thus easily driven to a central point, and are encompassed.

The guanacoes readily take to the water: several times at Port Valdes they were seen swimming from island to island. Byron, in his voyage, says he saw them drinking salt water. Some of our officers likewise saw a herd apparently drinking the briny fluid from a salina near Cape Blanco. I imagine in several parts of the country, if they do not drink salt water, they drink none at all. In the middle of the day, they frequently roll in the dust, in saucer-shaped hollows. The males fight together; two one day passed quite close to me, squealing and trying to bite each other; and several were shot with their hides deeply scored. Herds sometimes appear to set out on exploring-parties: at Bahia Blanca, where, within 30 miles of the coast, these animals are extremely unfrequent, I one day saw the tracks of thirty or forty, which had come in a direct line to a muddy salt-water creek. They then must have perceived that they were approaching the sea, for they had wheeled with the regularity of cavalry, and had returned back in as straight a line as they had advanced. The guanacoes have one singular habit, which is to me quite inexplicable; namely, that on successive days they drop their dung in the same defined heap. I saw one of these heaps which was eight feet in diameter, and necessarily was composed of a large quantity. Frezier remarks on this habit as common to the guanaco as well as to the llama; he says it is very useful to the Indians, who use the dung for fuel, and are thus saved the trouble of collecting it.

The guanacoes appear to have favourite spots for dying in. On the banks of the St Cruz, the ground was actually white with bones, in certain circumscribed spaces, which were generally bushy and all near the river. On one such spot I counted between ten and twenty heads. I particularly examined the bones; they did not appear, as some scattered ones which I had seen, gnawed or broken, as if dragged together by beasts of prey. The animals in most cases, must have crawled, before dying, beneath and amongst the bushes. Mr Bynoe informs me that during the last voyage, he observed the same circumstance on the banks of the Rio Gallegos. I do not at all understand the reason of this, but I may observe, that the wounded guanacoes at the St Cruz, invariably walked towards the river. At St Jago in the Cape de Verd islands I remember having seen in a retired ravine a corner under a cliff, where numerous goats' bones were collected: we at the time exclaimed, that it was the burial-ground of all the goats in the island. I mention these trifling circumstances, because in certain cases they might explain the occurrence of a number of uninjured bones in a cave, or buried under alluvial accumulations; and likewise the cause, why certain mammalia are more commonly embedded than others in sedimentary deposits. Any great flood of the St Cruz, would wash down many bones of the guanaco, but probably not a single one of the puma, ostrich, or fox. I may also observe, that almost every kind of waterfowl when wounded takes to the shore to die; so that the remains of birds, from this cause alone and independently of other reasons, would but rarely be preserved in a fossil state.

* * *

JANUARY 9TH, 1834 – Before it was dark the *Beagle* anchored in the fine spacious harbour of Port St Julian, situated about 110 miles to the south of Port Desire. On the south side of the harbour, a cliff of about 90 feet in height intersects a plain constituted of the formations above described; and its surface is strewed over with recent marine shells. The gravel, however, differently from that in every other locality, is covered by a very irregular and thin bed of a reddish loam, containing a few small calcareous concretions. The matter somewhat resembles that of the Pampas, and probably owes its origin either to a small stream having formerly entered the sea at that spot, or to a mud-bank similar to those now existing at the head of the harbour. In one spot this earthy matter filled up a hollow, or gully, worn quite through the gravel, and in this mass a group of large bones was embedded. The animal to which they belonged, must have lived, as in the case at Bahia Blanca, at a period long subsequent to the existence of the shells now inhabiting the coast. We may feel sure of this, because the formation of the lower terrace or plain, must necessarily have been posterior to those above it, and on the surface of the two higher ones, sea-shells of recent species are scattered. From the small physical change, which the last 100 feet elevation of the continent could have produced, the climate, as well as the general condition of Patagonia, probably was nearly the same, at the time when the animal was embedded, as it now is. This conclusion is moreover supported by the identity of the shells belonging to the two ages. Then immediately occurred the difficulty, how could any large quadruped have subsisted on these wretched deserts in lat. 49° 15'? I had no idea at the time, to what kind of animal these remains belonged. The puzzle, however, was soon solved when Mr Owen examined them; for he considers that they formed part of an animal allied to the guanaco or llama, but fully as large as the true camel. As all the existing members of the family of Camelidæ are inhabitants of the most sterile countries, so may we suppose was this extinct kind. The structure of the cervical vertebræ, the transverse processes not being perforated for the vertebral artery, indicates its affinity: some other parts, however, of its structure, probably are anomalous.

The most important result of this discovery, is the confirmation of the law that existing animals have a close relation in form with extinct species. As the guanaco is the characteristic quadruped of Patagonia, and the vicuna of the snow-clad summits of the Cordillera, so in bygone days, this gigantic species of the same family must have been conspicuous on the southern plains. We see this same relation of type between the existing and fossil Ctenomys, between the capybara (but less plainly, as shown by Mr Owen) and the gigantic Toxodon; and lastly, between the living and extinct Edentata. At the present day, in South America, there exist probably nineteen species of this order, distributed into several genera; while throughout the rest of the world there are but five. If, then, there is a relation between the living and the dead, we should expect that the Edentata would be numerous in the fossil state. I need only reply by enumerating the megatherium, and the three or four other great species, discovered at Bahia Blanca; the remains of some of which are also abundant over the whole immense territory of La Plata. I have already pointed out the singular relation between the armadilloes and their great prototypes, even in a point apparently of so little import-ance as their external covering.

The order of rodents at the present day, is most con-spicuous in South America, on account of the vast number and size of the species, and the multitude of individuals: according to the same law, we should expect to find their representatives in a fossil state. Mr Owen has shown how far the Toxodon is thus related; and it is moreover not improbable that another large animal has likewise a simi-lar affinity.

The teeth of the rodent nearly equalling in size those of the Capybara, which were discovered near Bahia Blanca, must also be remembered.

The law of the succession of types, although subject to some remarkable exceptions, must possess the highest interest to every philosophical naturalist, and was first clearly observed in regard to Australia, where fossil remains of a large and extinct species of Kangaroo and other marsupial animals were discovered buried in a cave. In America the most marked change among the mam-malia has been the loss of several species of Mastodon, of an elephant, and of the horse. These Pachydermata appear formerly to have had a range over the world, like that which deer and antelopes now hold. If Buffon had known of these gigantic armadilloes, llamas, great rodents, and lost pachydermata, he would have said with a greater semblance of truth, that the creative force in

America had lost its vigour, rather than that it had never possessed such powers.

It is impossible to reflect without the deepest astonishment, on the changed state of this continent. Formerly it must have swarmed with great monsters, like the southern parts of Africa, but now we find only the tapir, guanaco, armadillo, and capybara; mere pigmies compared to the antecedent races. The greater number, if not all, of these extinct quadrupeds lived at a very recent period; and many of them were contemporaries of the existing molluscs. Since their loss, no very great physical changes can have taken place in the nature of the country. What then has exterminated so many living creatures? In the Pampas, the great sepulchre of such remains, there are no signs of violence, but on the contrary, of the most quiet and scarcely sensible changes. At Bahia Blanca I endeavoured to show the probability that the ancient Edentata, like the present species, lived in a dry and sterile country, such as now is found in that neighbourhood. With respect to the camel-like llama of Patagonia, the same grounds which, before knowing more than the size of the remains, perplexed me, by not allowing any great change of climate, now that we can guess the habits of the animal, are strangely confirmed. What shall we say of the death of the fossil horse? Did those plains fail in pasture, which afterwards were overrun by thousands and tens of thousands of the successors of the fresh stock introduced with the Spanish colonist? In some countries, we may believe, that a number of species subsequently introduced, by consuming the food of the antecedent races, may have caused their extermination; but we can scarcely credit that the armadillo has devoured the food of the immense Megatherium, the capybara of the Toxodon, or the guanaco of the camel-like kind. But granting that all such changes have been small, yet we are so profoundly ignorant concerning the physiological relations, on which the life, and even health (as shown by epidemics) of any existing species depends, that we argue with still less safety about either the life or death of any extinct kind.

One is tempted to believe in such simple relations, as variation of climate and food, or introduction of enemies, or the increased numbers of other species, as the cause of the succession of races. But it may be asked whether it is probable that any such cause should have been in action during the same epoch over the whole northern hemisphere, so as to destroy the *Elephas primigenus*, on the shores of Spain, on the plains of Siberia, and in Northern America; and in a like manner, the *Bos urus*, over a range of scarcely less extent? Did such changes put a period to the life of *Mastodon angustidens*, and of the fossil horse, both in Europe and on the Eastern slope of the Cordillera in Southern America? If they did, they must have been changes common to the whole world; such as gradual refrigeration, whether from modifications of physical geography, or from central cooling. But on this assumption, we have to struggle with the difficulty that these supposed changes, although scarcely sufficient to affect molluscous animals either in Europe or South America, yet destroyed many quadrupeds in regions now characterized by *frigid, temperate*, and *warm* climates! These cases of extinction forcibly recall the idea (I do not wish to draw any close analogy) of certain fruit-trees, which, it has been asserted, though grafted on young stems, planted in varied situations, and fertilized by the richest manures, yet at one period, have all withered away and perished. A fixed and determined length of life has in such cases been given to thousands and thousands of buds (or individual germs), although produced in long succession. Among the greater number of animals, each individual appears nearly independent of its kind; yet all of one kind may be bound together by common laws, as well as a certain number of individual buds in the tree, or polypi in the Zoophyte.

I will add one other remark. We see that whole series of animals, which have been created with peculiar kinds of organization, are confined to certain areas; and we can hardly suppose these structures are only adaptations to peculiarities of climate or country; for otherwise, animals belonging to a distinct type, and introduced by man, would not succeed so admirably, even to the extermination of the aborigines. On such grounds it does not seem a necessary conclusion, that the extinction of species, more than their creation, should exclusively depend on the nature (altered by physical changes) of their country. All that at present can be said with certainty, is that, as with the individual, so with the species, the hour of life has run its course, and is spent.

* * *

APRIL 13TH – The *Beagle* anchored within the mouth of the Santa Cruz. This river is situated about 60 miles south of Port St Julian. During the last voyage, Captain Stokes proceeded 30 miles up, but then, from the want of provisions, was obliged to return. Excepting what was discovered at that time, scarcely any thing was known about this large river. Captain FitzRoy now determined to follow its course as far as time would allow. On the 18th, three whale-boats started, carrying three weeks' provisions; and the party consisted of twenty-five souls – a force which would have been sufficient to have defied a host of Indians. With a strong flood-tide, and a fine day, we made a good run, soon drank some of the fresh water, and were at night nearly above the tidal influence.

The river here assumed a size and appearance, which, even at the highest point we ultimately reached, was scarcely diminished. It was generally from 300 to 400 yards broad, and in the middle about 17 feet deep. The rapidity of the current, which in its whole course runs at the rate of from 4 to 6 knots an hour, is perhaps its most remarkable feature. The water is of a fine blue colour, but with a slight milky tinge, and not so transparent as at first sight would have been expected. It flows over a bed of pebbles, like those which compose the beach and surrounding plains. Although its course is winding, it runs through a valley which extends in a direct line to the westward. This valley varies from 5 to 10 miles in breadth; it is bounded by step-formed terraces, which rise in most parts one above the other to the height of 500 feet, and have on the opposite sides a remarkable correspondence.

APRIL 19TH – Against so strong a current, it was of course quite impossible to row or sail. Consequently the three boats were fastened together head and stern, two hands left in each, and the rest came on shore to track. As the general arrangements, made by Captain FitzRoy, were very good for facilitating the work of all, and as all had a share of it, I will describe the system. The party, including everyone, was divided into two spells, each of which hauled at the tracking line alternately for an hour and a half. The officers of each boat lived with, ate the same food, and slept in the same tent with their crew, so that each boat was quite independent of the others. After sunset, the first level spot where any bushes were growing, was chosen for our night's lodging. Each of the crew took it in turns to be cook. Immediately the boat was hauled up, the cook made his fire; two others pitched the tent; the coxswain handed the things out of the boat; the rest carried them up to the tents, and collected firewood. By this order, in half an hour, every thing was ready for the night. A watch of two men and an officer was always kept, whose duty it was to look after the boats, keep up the fire, and guard against Indians. Each in the party had his one hour every night.

During this day we tracked but a short distance, for there were many islets, covered by thorny bushes, and the channels between them were shallow.

APRIL 20TH – We passed the islands and set to work. Our regular day's march, although it was hard enough, carried us on an average only 10 miles in a straight line, and perhaps 15 or 20 altogether. Beyond the place where we slept last night the country is completely *terra incognita*, for it was there that Captain Stokes turned back. We saw in the distance a great smoke, and found the skeleton of a horse, so we knew that Indians were in the neighbourhood. On the next morning (21st) tracks of a party of horse, and marks left by the trailing of the *chuzos* were observed on the ground. It was generally thought they must have reconnoitred us during the night. Shortly afterwards we came to a spot, where from the fresh footsteps of men, children, and horses, it was evident the party had crossed the river.

* * *

APRIL 29TH – From some high land we hailed with joy the white summits of the Cordillera, as they were seen occasionally peeping through their dusky envelope of clouds. During the few succeeding days, we continued to get on slowly, for we found the river-course very tortuous, and strewed with immense fragments of various ancient slaty rocks, and of granite. The plain bordering the valley had here attained an elevation of about 1,100 feet, and its character was much altered. The well-rounded pebbles of porphyry were in this part mingled with many immense angular fragments of basalt and of the rocks above mentioned. The first of these erratic blocks which I noticed, was 67 miles distant from the nearest mountain; another which had been transported to rather a less distance, measured 5 yards square, and projected 5 feet above the gravel. Its edges were so angular, and its size so great, that I at first mistook it for a rock *in situ*, and took out my compass to observe the direction of its cleavage. The plains here were not quite so level as those nearer the coast, but yet, they betrayed little signs of any violent action. Under these circumstances, it would be difficult, as it appears to me, to explain this phenomenon on any theory, excepting through that of transport by ice while the country was under water. But this is a subject to which I shall again recur.

During the two last days we met with signs of horses, and with several small articles which had belonged to the Indians, – such as parts of a mantle and a bunch of ostrich feathers – but they appeared to have been lying long on the ground. Between the place where the Indians had so lately crossed the river and this neighbourhood, though so many miles apart, the country appears to be quite unfrequented. At first, considering the abundance of the guanacoes, I was surprised at this; but it is explained by the stony nature of the plains, which would soon disable an unshod horse from taking part in the chase. Nevertheless, in two places in this very central region, I found small heaps of stones, which I do not think could have been accidentally thrown together. They were placed on points, projecting over the edge of the highest lava cliff, and they resembled, but on a small scale, those near Port Desire.

MAY 4TH – Captain FitzRoy determined to take the boats no higher. The river had a winding course, and was very rapid; and the appearance of the country offered no temptation to proceed any further. Every where we met with the same productions, and the same dreary landscape. We were now 140 miles distant from the Atlantic, and about 60 from the nearest arm of the Pacific. The valley in this upper part expanded into a wide basin, bounded on the north and south by the basaltic platforms, and fronted by the long range of the snow-clad Cordillera. But we viewed these grand mountains with regret, for we were obliged to imagine their form and nature, instead of standing, as we had hoped, on their crest, and looking down on the plain below. Besides the useless loss of time which an attempt to ascend any higher would have cost us, we had already been for some days on half allowance of bread. This, although really enough for any reasonable men, was, after our hard day's march, rather scanty food. Let those alone who have never tried it, exclaim about the comfort of a light stomach and an easy digestion.

5TH – Before sunrise we commenced our descent. We shot down the stream with great rapidity, generally at the rate of 10 knots an hour. In this one day we effected what had cost us five-and-a-half hard days' labour in ascending. On the 8th, we reached the *Beagle* after our twenty-one days' expedition. Every one excepting myself had cause to be dissatisfied; but to me the ascent afforded a most interesting section of the great tertiary formation of Patagonia.

Tierra del Fuego

DECEMBER 17TH, 1832 – Having now finished with Patagonia, I will describe our first arrival in Tierra del Fuego. A little after noon we doubled Cape St Diego, and entered the famous strait of Le Maire. We kept close to the Fuegian shore, but the outline of the rugged, inhospitable Staten land was visible amidst the clouds. In the afternoon we anchored in the Bay of Good Success. While entering we were saluted in a manner becoming the inhabitants of this savage land. A group of Fuegians partly concealed by the entangled forest, were perched on a wild point overhanging the sea; and as we passed by, they sprang up, and waving their tattered cloaks sent forth a loud and sonorous shout. The savages followed the ship, and just before dark we saw their fire, and again heard their wild cry. The harbour consists of a fine piece of water half surrounded by low rounded mountains of clay-slate, which are covered to the water's edge by one dense gloomy forest. A single glance at the landscape was sufficient to show me, how widely different it was from any thing I had ever beheld. At night it blew a gale of wind, and heavy squalls from the mountains swept past us. It would have been a bad time out at sea, and we, as well as others, may call this Good Success Bay.

In the morning, the Captain sent a party to communicate with the Fuegians. When we came within hail, one of the four natives who were present advanced to receive us, and began to shout most vehemently, wishing to direct us where to land. When we were on shore the party looked rather alarmed, but continued talking and making gestures with great rapidity. It was without exception the most curious and interesting spectacle I had ever beheld. I could not have believed how wide was the difference, between savage and civilized man. It is greater than between a wild and domesticated animal, in as much as in man there is a greater power of improvement. The chief spokesman was old, and appeared to be the head of the family; the three others were powerful young men, about 6 feet high. The women and children had been sent away. These Fuegians are a very different race from the stunted miserable wretches further to the westward. They are much superior in person, and seem closely allied to the famous Patagonians of the Strait of Magellan. Their only garment consists of a mantle made of guanaco skin, with the wool outside; this they wear just thrown over their shoulders, as often leaving their persons exposed as covered. Their skin is of a dirty coppery red colour.

The old man had a fillet of white feathers tied round his head, which partly confined his black, coarse, and entangled hair. His face was crossed by two broad transverse bars; one painted bright red reached from ear to ear, and included the upper lip; the other, white like chalk, extended parallel and above the first, so that even his eyelids were thus coloured. Some of the other men were ornamented by streaks of black powder, made of charcoal. The party altogether closely resembled the devils which come on the stage in such plays as Der Freischutz.

Their very attitudes were abject, and the expression of their countenances distrustful, surprised, and startled. After we had presented them with some scarlet cloth, which they immediately tied round their necks, they became good friends. This was shown by the old man patting our breasts, and making a chuckling kind of noise, as people do when feeding chickens. I walked with the old man, and this demonstration of friendship was repeated several times; it was concluded by three hard slaps, which were given me on the breast and back at the same time. He then bared his bosom for me to return the compliment, which being done, he seemed highly pleased. The language of these people, according to our notions, scarcely deserves to be called articulate. Captain Cook has compared it to a man clearing his throat, but certainly no European ever cleared his throat with so many hoarse, guttural, and clicking sounds.

They are excellent mimics: as often as we coughed or yawned, or made any odd motion, they immediately imitated us. Some of our party began to squint and look awry; but one of the young Fuegians (whose whole face was painted black, excepting a white band across his eyes) succeeded in making far more hideous grimaces. They could repeat with perfect correctness, each word in any sentence we addressed them, and they remembered such words for some time. Yet we Europeans all know how difficult it is to distinguish apart the sounds in a foreign language. Which of us, for instance, could follow an American Indian through a sentence of more than three words? All savages appear to possess, to an uncommon degree, this power of mimicry. I was told almost in the same words, of the same ludicrous habits among the Caffres: the Australians, likewise, have long been notorious for being able to imitate and describe the gait of any man, so that he may be recognized. How can this faculty be explained? is it a consequence of the more practised habits of perception and keener senses, common to all men in a savage state, as compared to those long civilized?

When a song was struck up by our party, I thought the Fuegians would have fallen down with astonishment. With equal surprise they viewed our dancing; but one of the young men, when asked, had no objection to a little waltzing. Little accustomed to Europeans as they appeared to be, yet they knew, and dreaded our fire-arms; nothing would tempt them to take a gun in their hands. They begged for knives, calling them by the Spanish word 'cuchilla'. They explained also what they wanted, by acting as if they had a piece of blubber in their mouth, and then pretending to cut instead of tear it.

It was interesting to watch the conduct of these people towards Jemmy Button (one of the Fuegians who had been taken, during the former voyage, to England): they immediately perceived the difference between him and the rest, and held much conversation between themselves on the subject. The old man addressed a long harangue to Jemmy, which it seems was to invite him to stay with them. But Jemmy understood very little of their language, and was, moreover, thoroughly ashamed of his country-men. When York Minster (another of these men) came on shore, they noticed him in the same way, and told him he ought to shave; yet he had not twenty dwarf hairs on his face, whilst we all wore our untrimmed beards. They examined the colour of his skin, and compared it with ours. One of our arms being bared, they expressed the liveliest surprise and admiration at its whiteness. We thought that they mistook two or three of the officers, who were rather shorter and fairer (though adorned with large beards), for the ladies of our party. The tallest amongst the Fuegians was evidently much pleased at his height being noticed. When placed back to back with the tallest of the boat's crew, he tried his best to edge on higher ground, and to stand on tiptoe. He opened his mouth to show his teeth, and turned his face for a side view; and all this was done with such alacrity, that I dare say he thought himself the handsomest man in Tierra del

Fuego. After the first feeling on our part of grave astonishment was over, nothing could be more ludicrous or
interesting than the odd mixture of surprise and imitation
which these savages every moment exhibited.

The next day I attempted to penetrate some way into
the country. Tierra del Fuego may be described as a
mountainous country, partly submerged in the sea, so
that deep islets and bays occupy the place where valleys
should exist. The mountain sides (except on the exposed
western coast) are covered from the water's edge upwards
by one great forest. The trees reach to an elevation of
between 1,000 and 1,500 feet; and are succeeded by a
band of peat, with minute alpine plants; and this again
is succeeded by the line of perpetual snow, which, according to Captain King, in the Strait of Magellan descends
to between 3,000 and 4,000 feet. To find an acre of level
land in any part of the country is most rare. I recollect
only one little flat near Port Famine, and another of rather
larger extent near Goeree Road. In both these cases, and
in all others, the surface was covered by a thick bed of
swampy peat. Even within the forest the ground is concealed by a mass of slowly putrefying vegetable matter,
which, from being soaked with water, yields to the foot.

Finding it nearly hopeless to push my way through the
wood, I followed the course of a mountain torrent. At
first, from the waterfalls and number of dead trees, I
could hardly crawl along; but the bed of the stream soon
became a little more open, from the floods having swept

the sides. I continued slowly to advance for an hour along the broken and rocky banks; and was amply repaid by the grandeur of the scene. The gloomy depth of the ravine well accorded with the universal signs of violence. On every side were lying irregular masses of rock and up-torn trees; other trees, though still erect, were decayed to the heart and ready to fall. The entangled mass of the thriving and the fallen reminded me of the forests within the tropics; – yet there was a difference; for in these still solitudes, Death, instead of Life, seemed the predominant spirit. I followed the water-course till I came to a spot where a great slip had cleared a straight space down the mountain side. By this road I ascended to a considerable elevation, and obtained a good view of the surrounding woods. The trees all belong to one kind, the *Fagus betuloides*, for the number of the other species of beech, and of the Winter's bark, is quite inconsiderable. This tree keeps its leaves throughout the year; but its foliage is of a peculiar brownish-green colour, with a tinge of yellow. As the whole landscape is thus coloured, it has a sombre, dull appearance; nor is it often enlivened by the rays of the sun.

Strait of Magellan

DECEMBER 21ST – The *Beagle* got under way: and on the succeeding day, favoured to an uncommon degree by a fine easterly breeze, we closed in with the Barnevelts, and, running past Cape Deceit with its stony peaks, about three o'clock doubled the weather beaten Cape Horn! The evening was calm and bright and means of judging of the distance, how the mountain appeared to rise in height.

The Fuegians twice came and plagued us. As there were many instruments, clothes, and men on shore, it was thought necessary to frighten them away. The first time, a few great guns were fired, when they were far distant. It was most ludicrous to watch through a glass the Indians, as often as the shot struck the water, take up stones, and as a bold defiance, throw them towards the ship, though about a mile and a half distant! A boat was then sent with orders to fire a few musket-shot wide of them. The Fuegians hid themselves behind the trees; and for every discharge of the musket they fired their arrows: all, however, fell short of the boat, and the officer as he

pointed at them laughed. This made the Fuegians frantic
with passion, and they shook their mantles in vain rage.
At last seeing the balls cut and strike the trees, they ran
away; and we were left in peace and quietness.

On a former occasion, when the *Beagle* was here in the
month of February, I started one morning at four o'clock
to ascend Mount Tarn, which is 2,600 feet high, and is
the most elevated point in this immediate neighbourhood. We went in a boat to the foot of the mountain (but
not to the best part), and then began our ascent. The
forest commences at the line of high-water mark, and
during the two first hours I gave over all hopes of reach-
ing the summit. So thick was the wood, that it was
necessary to have constant recourse to the compass; for
every landmark, though in a mountainous country, was
completely shut out. In the deep ravines, the death-like
scene of desolation exceeded all description; outside it
was blowing a gale, but in these hollows, not even a
breath of wind stirred the leaves of the tallest trees. So
gloomy, cold, and wet was every part, that not even the
fungi, mosses, or ferns, could flourish. In the valleys it
was scarcely possible to crawl along, they were so completely barricaded by the great mouldering trunks, which
had fallen down in every direction. When passing over
these natural bridges, one's course was often arrested by
sinking knee deep into the rotten wood; at other times,
when attempting to lean against a firm tree, one was startled by finding a mass of decayed matter ready to fall at

the slightest touch. We at last found ourselves among the stunted trees, and then soon reached the bare ridge, which conducted us to the summit. Here was a view characteristic of Tierra del Fuego; – irregular chains of hills, mottled with patches of snow, deep yellowish-green valleys, and arms of the sea intersecting the land in many directions. The strong wind was piercingly cold, and the atmosphere rather hazy, so that we did not stay long on the top of the mountain. Our descent was not quite so laborious as our ascent; for the weight of the body forced a passage, and all the slips and falls were in the right direction.

* * *

The perfect preservation of the Siberian animals, perhaps presented, till within a few years, one of the most difficult problems which geology ever attempted to solve. On the one hand it was granted, that the carcasses had not been drifted from any great distance by any tumultuous deluge, and on the other it was assumed as certain, that when the animals lived, the climate must have been so totally different, that the presence of ice in the vicinity was as incredible, as would be the freezing of the Ganges. Mr Lyell in his *Principles of Geology* has thrown the greatest light on this subject, by indicating the northerly course of the existing rivers with the probability that they formerly carried carcasses in the same direction; by showing (from Humboldt) how far the inhabitants of the hottest countries sometimes wander; by insisting on the caution necessary in judging of habits between animals of the same genus, when the species are not identical; and especially by bringing forward in the clearest manner the probable change from an insular to an extreme climate, as the consequence of the elevation of the land, of which proofs have lately been brought to light.

In a former part of this volume, I have endeavoured to prove, that as far as regards the *quantity* of food, there is no difficulty in supposing that these large quadrupeds inhabited sterile regions, producing but a scanty vegetation. With respect to temperature, the woolly covering both of the elephant and the rhinoceros seems at once to render it at least probable (although it has been argued that some animals living in the hottest regions are thickly clothed) that they were fitted for a cold climate. I suppose no reason can be assigned why, during a former epoch, when the pachydermata abounded over the greater part of the world, some species should not have been fitted for the northern regions, precisely as now happens with deer and several other animals. If, then, we believe that the climate of Siberia, anteriorly to the physical changes above alluded to, had some resemblance with that of the southern hemisphere at the present day – a circumstance which harmonizes well with other facts, as I think has been shown by the imaginary case, when we transported existing phenomena from one to the other hemisphere – the following conclusions may be deduced as probable: first, that the degree of cold formerly was not excessive; secondly, that snow did not for a long time together cover the ground (such not being the case at the extreme parts 55°–56° of S. America); thirdly, that the vegetation partook of a more tropical character than it now does in the same latitudes; and lastly, that at but a short distance to the northward of the country thus circumstanced (even not so far as where Pallas found the entire rhinoceros), the soil might be perpetually congealed: so that if the carcass of any animal should once be buried a few feet beneath the surface, it would be preserved for centuries.

Both Humboldt and Lyell have remarked, that at the present day, the bodies of any animals, wandering beyond the line of perpetual congelation which extends as far south as 62°, if once embedded by any accident a few feet beneath the surface, would be preserved for an indefinite length of time: the same would happen with carcasses drifted by the rivers; and by such means the extinct mammalia may have been entombed. There is only one small step wanting, as it appears to me, and the whole problem would be solved with a degree of simplicity very striking, compared with the several theories first invented. From the account given by Mr Lyell of the Siberian plains, with their innumerable fossil bones, the relics of many successive generations, there can be little doubt that the beds were accumulated either in a shallow sea, or in an estuary. From the description given in Beechey's voyage of

Eschscholtz Bay, the same remark is applicable to the north-west coast of America: the formation there appears identical with the common littoral deposits recently elevated, which I have seen on the shores of the southern part of the same continent. It seems also well established, that the Siberian remains are only exposed where the rivers intersect the plain. With this fact, and the proofs of recent elevation, the whole case appears to be precisely similar to that of the Pampas: namely, that the carcasses were formerly floated into the sea, and the remains covered up in the deposits which were then accumulating. These beds have since been elevated; and as the rivers excavate their channels the entombed skeletons are exposed.

Here then, is the difficulty: how were the carcasses preserved at the bottom of the sea? I do not think it has been sufficiently noticed, that the preservation of the animal with its flesh was an occasional event, and not directly consequent on its position far northward. Cuvier refers to the voyage of Billing as showing that the *bones* of the elephant, buffalo, and rhinoceros, are nowhere so abundant as on the islands between the mouths of the Lena and Indigirska. It is even said that excepting some hills of rock, the whole is composed of sand, ice, and bones. These islands lie to the northward of the place where Adams found the mammoth with its flesh preserved, and even 10° north of the Wiljui, where the rhinoceros was discovered in a like condition. In the case of the *bones* we may suppose that the carcasses were drifted into a deeper sea, and there remaining at the bottom, the flesh decomposed. But in the second and more extraordinary case, where putrefaction seems to have been arrested, the body probably was soon covered up by deposits which were then accumulating. It may be asked, whether the mud a few feet deep, at the bottom of a shallow sea which is annually frozen, has a temperature higher than 32°? It must be remembered how intense a degree of cold is required to freeze salt water; and that the mud at some depth below the surface, would have a low mean temperature, precisely in the same manner as the subsoil on the land is frozen in countries which enjoy a short but hot summer. If this be possible, the entombment of these extinct quadrupeds is rendered very simple; and with regard to the conditions of their former existence, the principal difficulties have, I think, already been removed.

* * *

There is one marine production, which from its importance is worthy of a particular history. It is the kelp or *Fucus giganteus* of Solander. This plant grows on every rock from low-water mark to a great depth, both on the outer coast and within the channels. I believe, during the voyages of the *Adventure* and *Beagle*, not one rock near the surface was discovered, which was not buoyed by this floating weed. The good service it thus affords to vessels navigating near this stormy land is evident; and it certainly has saved many a one from being wrecked. I know few things more surprising than to see this plant growing and flourishing amidst those great breakers of the western ocean, which no mass of rock, let it be ever so hard, can long resist. The stem is round, slimy, and smooth, and seldom has a diameter of so much as an inch. A few taken together are sufficiently strong to support the weight of the large loose stones to which in the inland channels they grow attached; and some of these stones are so heavy, that when drawn to the surface they can scarcely be lifted into a boat by one person.

Captain Cook, in his second voyage, says, that at Kerguelen Land 'some of this weed is of a most enormous length, though the stem is not much thicker than a man's thumb. I have mentioned, that on some of the shoals upon which it grows, we did not strike ground with a line of 24 fathoms. The depth of water, therefore, must have been greater. And as this weed does not grow in a perpendicular direction, but makes a very acute angle with the bottom, and much of it afterwards spreads many fathoms on the surface of the sea, I am well warranted to say that some of it grows to the length of sixty fathoms and upwards.' Certainly at the Falkland Islands, and about Tierra del Fuego, extensive beds frequently spring up from 10- and 15-fathom water. I do not suppose the stem of any other plant attains so great a length as 360 feet, as stated by Captain Cook. Its geographical range is very considerable; it is found from the extreme southern islets near Cape Horn, as far north, on the eastern coast (according to information given me by Mr Stokes), as lat. 43° – and on the western it was tolerably abundant, but far from luxuriant, at Chiloe, in lat. 42°. It may possibly extend a little further northward, but is soon succeeded by a different species. We thus have a range of 15° in latitude; and as Cook, who must have been well acquainted with the species, found it at Kerguelen Land, no less than 140° in longitude.

The number of living creatures of all orders, whose existence intimately depends on the kelp, is wonderful. A great volume might be written, describing the inhabitants of one of these beds of sea-weed. Almost every leaf, excepting those that float on the surface, is so thickly incrusted with coral-lines, as to be of a white colour. We find exquisitely-delicate structures, some inhabited by simple hydra-like polypi, others by more organized kinds, and beautiful compound Ascidiæ. On the flat surfaces of the leaves various patelliform shells, Trochi, uncovered molluscs, and some bivalves are attached. Innumerable crustacea frequent every part of the plant. On shaking the great entangled roots, a pile of small fish, shells, cuttle-fish, crabs of all orders, sea-eggs, star-fish, beautiful Holuthuriæ (some taking the external form of the nudibranch molluscs), Planariæ, and crawling nereidous animals of a multitude of forms, all fall out together. Often as I recurred to a branch of the kelp, I never failed to discover animals of new and curious structures. In Chiloe, where, as I have said, the kelp did not thrive very well, the numerous shells, coral-lines, and crustacea were absent; but there yet remained a few of the flustraceæ, and some compound Ascidiæ; the latter, however, were of different species from those in Tierra del Fuego. We here see the fucus possessing a wider range than the animals which use it as an abode.

I can only compare these great aquatic forests of the southern hemisphere with the terrestrial ones in the intertropical regions. Yet if the latter should be destroyed in any country, I do not believe nearly so many species of animals would perish, as, under similar circumstances, would happen with the kelp. Amidst the leaves of this plant numerous species of fish live, which nowhere else would find food or shelter; with their destruction the many cormorants, divers, and other fishing birds, the otters, seals, and porpoises, would soon perish also; and lastly, the Fuegian savage, the miserable lord of this miserable land, would redouble his cannibal feast, decrease in numbers, and perhaps cease to exist.

JUNE 8TH – We weighed anchor early in the morning, and left Port Famine. Captain FitzRoy determined to leave the Strait of Magellan by the Magdalen channel, which had not long been discovered. Our course lay due south, down that gloomy passage which I have before alluded to, as appearing to lead to another and worse world. The wind was fair, but the atmosphere was very thick; so that we missed much curious scenery. The dark ragged clouds were rapidly driven over the mountains, from their summits nearly to their bases. The glimpses which we caught through the dusky mass were highly interesting: jagged points, cones of snow, blue glaciers, strong outlines marked on a lurid sky, were seen at different distances and heights. In the midst of such scenery we anchored at Cape Turn, close to Mount Sarmiento, which was then hidden in the clouds. At the base of the lofty and almost perpendicular sides of our little cove, there was one deserted wigwam, and it alone reminded us that man sometimes wandered in these desolate regions. But it would be difficult to imagine a scene where he seemed to have less claims, or less authority. The inanimate works of nature – rock, ice, snow, wind, and water – all warring with each other, yet combined against man – here reigned in absolute sovereignty.

JUNE 9TH – In the morning we were delighted by seeing the veil of mist gradually rise from Sarmiento, and display it to our view. This mountain, which is one of the highest in Tierra del Fuego, has an elevation of 6,800 feet. Its base, for about an eighth of its total height, is clothed by dusky woods, and above this a field of snow extends to the summit. These vast piles of snow, which never melt, and seem destined to last as long as the world holds together, present a noble and even sublime spectacle. The outline of the mountain was admirably clear and defined. Owing to the abundance of light reflected from the white and glittering surface, no shadows are cast on any part; and those lines which intersect the sky can alone be distinguished: hence the mass stood out in the boldest relief. Several glaciers descended in a winding course, from the snow to the sea-coast: they may be likened to great frozen Niagaras; and perhaps these cataracts of blue ice are to the full as beautiful as the moving ones of water. By night we reached the western part of the channel; but the water was so deep that no anchorage could be found. We were in consequence obliged to stand off and on, in this narrow arm of the sea, during a pitch-dark night of fourteen hours long.

JUNE 10TH – In the morning we made the best of our way into the open Pacific. The Western coast generally consists of low, rounded, quite barren, hills of granite and greenstone. Sir John Narborough called one part South Desolation, because it is 'so desolate a land to behold'; and well indeed might he say so. Outside the main islands there are numberless scattered rocks, on which the long swell of the open ocean incessantly rages. We passed out between the East and West Furies, and a little further northward there are so many breakers that the sea is called the Milky Way. One sight of such a coast is enough to make a landsman dream for a week about shipwreck, peril, and death; and with this sight, we bade farewell for ever to Tierra del Fuego.